Nature's Habitats

IN THE OCEANS AND SEAS

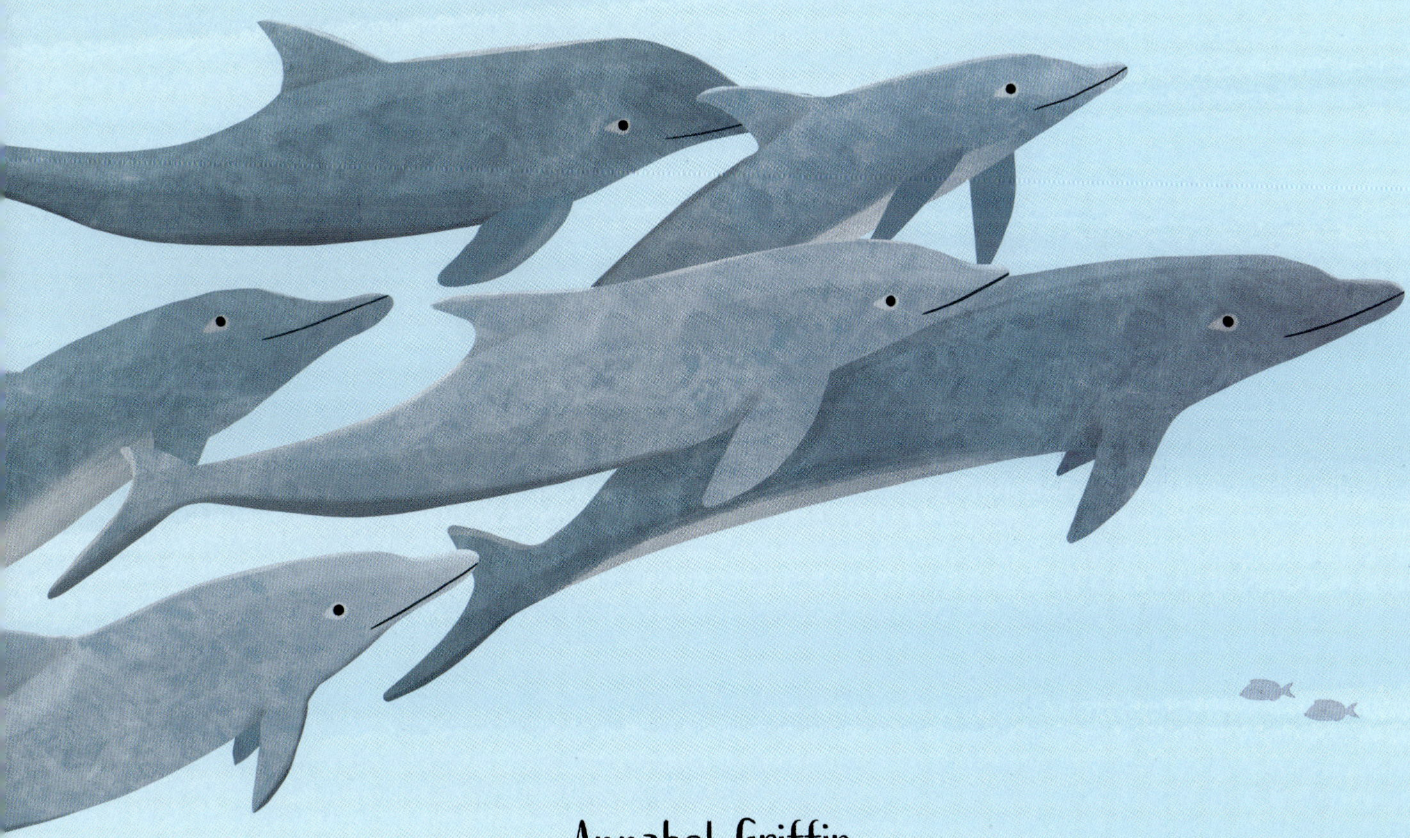

Annabel Griffin
Illustrated by Rose Maclachlan

First published in Great Britain in 2024
by Hungry Tomato Ltd
F15, Old Bakery Studios,
Blewetts Wharf, Malpas Road
Truro, Cornwall, TR1 1QH, UK

Copyright© 2024 Hungry Tomato Ltd

No part of this publication may be reproduced, stored in a retrieval system, or transmitted in any form or by any means, electronic, mechanical, photocopying, recording, or otherwise, without prior written permission of the copyright owner.

A CIP catalogue record for this book is available from the British Library.

ISBN 9781835693513

Printed in China

Discover more at:
www.hungrytomato.com

Psst! I'm hiding on every page. Can you spot me?

CONTENTS

> Words in bold capital letters **LIKE THIS** can be found in the glossary.

In the Oceans and Seas	4
Here Come the Sharks!	6
Life on the Reef	8
Weird and Wonderful Fish	10
Marine Mammals	12
I See Shells!	14
But Is It a Fish?	16
Monsters of the Deep	18
How Deep is the Ocean?	20
Did You Know?	22
Who was Hiding?	23
Glossary	24

IN THE OCEANS AND SEAS

There is so much to see in our oceans and seas! Can you spot the underwater creatures that make their homes here?

HERE COME THE SHARKS!

A shark is a type of fish. There are over 500 different kinds of shark! These are just a few of them.

Killing machines
Great whites are deadly PREDATORS but it's very rare for them to attack humans.

Great White Shark

Gentle giants
Whale sharks are the biggest fish in the sea and can live to be 150 years old!

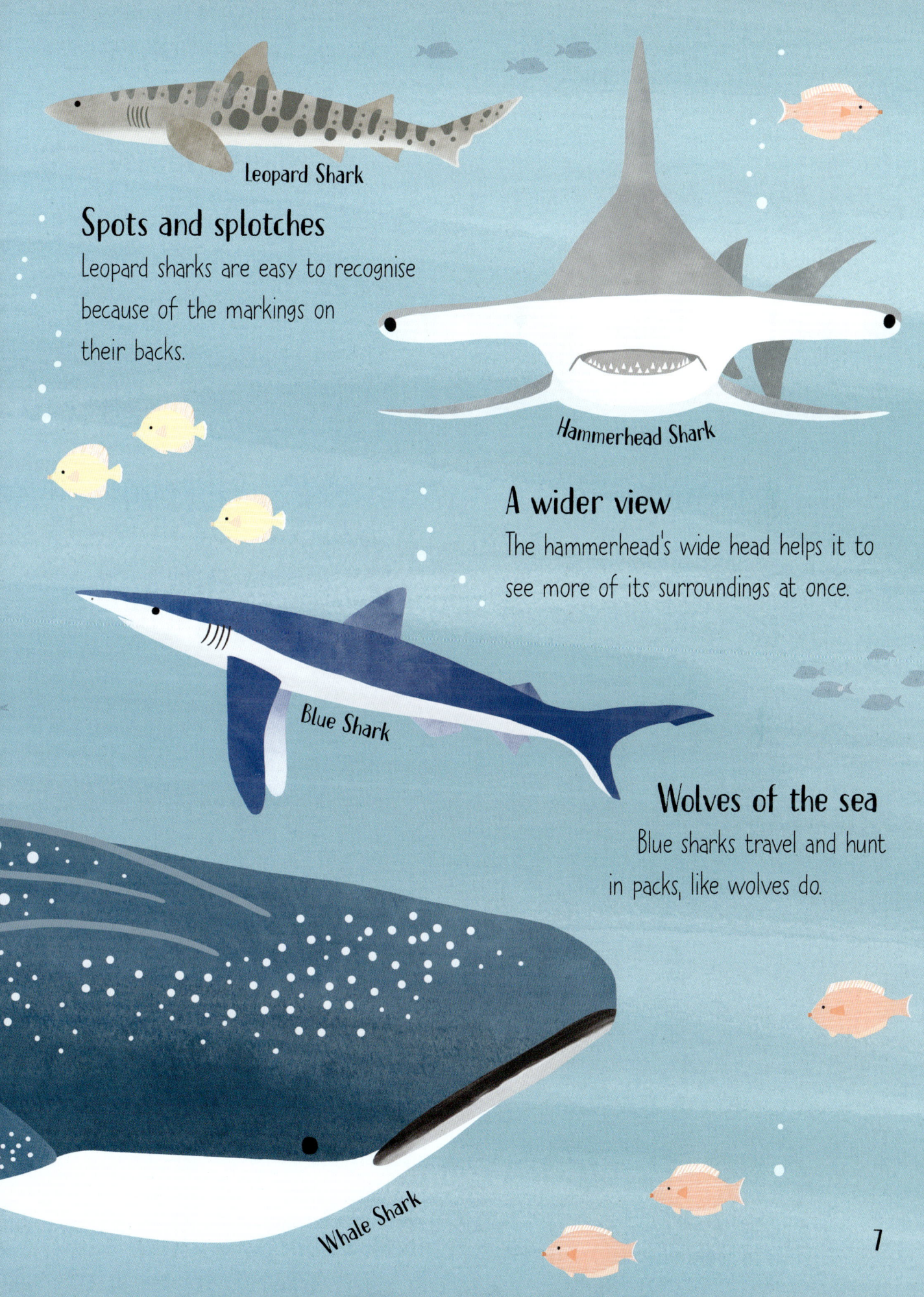

Leopard Shark

Spots and splotches
Leopard sharks are easy to recognise because of the markings on their backs.

Hammerhead Shark

A wider view
The hammerhead's wide head helps it to see more of its surroundings at once.

Blue Shark

Wolves of the sea
Blue sharks travel and hunt in packs, like wolves do.

Whale Shark

LIFE ON THE REEF

Coral reefs are large underwater structures made of coral. They are home to thousands of sea creatures!

Slithery swimmers
Sea snakes are some of the deadliest snakes in the world.

An unusual friendship
Sea anemones have poisonous stinging tentacles to catch fish but...

...clownfish are **IMMUNE** to their sting. They make anemones their home and help to attract other fish for them to eat.

Hawksbill Sea Turtle

Here for the food!
Sponges are the Hawksbill turtle's favourite things to eat.

Sea fans are a type of coral.

Sea Fan

Plant or animal?
Coral, sponges, anemones and urchins may look like weird plants, but they are actually all animals!

Sea Urchin

WEIRD AND WONDERFUL FISH

There are some very strange looking fish in the ocean! Some of them have special skills too.

Big and brainy
Manta rays are the largest rays in the world and have the biggest brains of any fish. That's smart!

Manta Ray

All puffed up
When they get scared, pufferfish INFLATE to several times their normal size; like a water balloon.

Pufferfish

World's fastest fish
Sailfish are the fastest fish in the ocean, and can reach speeds of up to 68 miles per hour!

Masters of disguise
Stonefish are CAMOUFLAGED to look like stones and coral on the seabed.

Dangerous beauty
The lionfish may look pretty but it is covered in very sharp, VENOMOUS fins.

MARINE MAMMALS

It's not just fish in the oceans and sea, there are plenty of MAMMALS too! Mammals don't have GILLS, so they have to come to the surface of the water to breathe.

Bottlenose Dolphins

Humpback Whale

Songs of the sea
Humpback whales are famous for singing songs to each other.

Blue Whale

The biggest ever!
The blue whale is the largest known animal to have ever existed! They can weigh as much as 40 African elephants!

On land and on sea
Unlike whales and dolphins, seals can live on land too.

Sticking together
Dolphins travel together in groups called *pods*.

Sea cows
Baby manatees, known as calves, will stay close to their mums for up to two years.

Hawaiian Monk Seal

Manatees

I SEE SHELLS!

Where do seashells come from? Our oceans and seas of course! Lots of creatures have shells. Here are some for you to spot.

Lots of legs
All crabs have 10 legs. Their front legs have claws which they use to fight with and catch their food.

Crab

Giant Clam

Would you hide inside?
Giant clam shells are often big enough that you could fit inside them!

A living fossil
Nautiluses were living in the sea 265 million years before dinosaurs existed! That makes them living **FOSSILS**!

Nautilus

Hidden chompers
Did you know a lobster's teeth aren't in its mouth? They are in its stomach!

Lobster

Plenty of snails
Many seashells you'll find on the beach will have belonged to sea snails. They come in lots of different colours, shapes and sizes.

Sea Snail

BUT IS IT A FISH?

Sometimes things are not what they seem and names can be misleading. Can you tell which of these creatures are fish and which are something else?

Slip and slide
Eels may look more like snakes but they are a type of **fish**.

Moray Eel

Super stars
Despite their name, starfish are **not** actually fish. They are related to the same marine animal as sea urchins.

Starfish

Jiggling jellies
These strange and beautiful creatures are **not** fish. They are related to coral and sea anemones.

Horsing around
You may not think it, but seahorses are **fish**. They are definitely not horses!

Eight-legged brainbox
Octopuses are highly intelligent beings, but they are **not** fish. They have more in common with slugs and snails.

MONSTERS OF THE DEEP

In the deepest, darkest parts of the ocean live some of the scariest and strangest sea creatures of all!

Eyes up!
This weird looking fish has a see-through head! Its eyes are actually on the inside of its head.

Barreleye Fish

Caped creature
The vampire squid isn't really a squid... or a vampire! They are related to octopuses and have eight arms connected to their "cape".

Vampire Squid

Nightmare fish

With their giant fangs, these fish may look terrifying, but they are quite small and harmless to humans.

Fangtooth

Long-nosed Chimaera

BOO!

These spooky fish are also known as ghost sharks.

Night fishing

This freaky fish has a glowing rod that sticks out of its head to attract PREY.

Deep Sea Anglerfish

HOW DEEP IS THE OCEAN?

The deeper you go in the ocean, the less sunlight there is, so it gets darker and colder. The ocean can be split into zones, according to the amount of light there is. Different animals and plants live in each zone, but the deeper you go, the fewer you'll find.

sea level

sunlight zone

200 metres

DID YOU KNOW?

Dads with a difference
Male seahorses get pregnant and give birth instead of the females.

Interesting family
Manatees are known as 'sea cows' but they are really related to elephants!

Magic stars
If a starfish loses a leg, it can grow it back!

WHO WAS HIDING?

Did you spot the little sea turtle playing hide-and-seek in each ocean scene?

This is the Kemp's ridley sea turtle. It is the smallest type of sea turtle there is. Adult turtles are normally between 58-71 cm long.

My favourite food is crab!

You would have a hard time finding these little creatures in real life. They are the world's most **ENDANGERED** sea turtle.

Couldn't find it in the deep ocean on pages 18-19? That's because it is much too deep and dark for turtles down there!

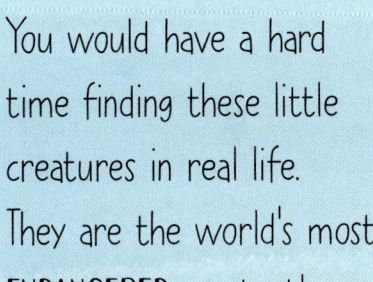

Help!

GLOSSARY

camouflaged - to look like something else so as not to be easily seen.

endangered - if a type of animal or plant is in danger of dying out forever, then they are known as endangered.

fossils - the remains or traces of plants and animals that lived a very long time ago.

gills - a body part that fish and some other animals use to breathe underwater.

immune - to be protected from/unaffected by something, such as an illness or a poison.

inflate - to make larger.

mammals - are animals with specific features. They all have hair or fur, drink milk from their mothers as babies, have a backbone, and are warm-blooded (they can keep their bodies warm, even when it's cold outside). Humans are mammals too!

predators - animals that hunt and kill other animals for food.

prey - an animal that is hunted by other animals for food.

venomous - poisonous, or containing poison.

The Author
Annabel Griffin is a writer and artist based in London, UK. Having worked as a bookseller for many years, she is now working in the children's publishing industry. Annabel's most recent publications include *Seasons* and *The Spectacular Lives of Sharks*.

The Illustrator
Rose Maclachlan is an illustrator based in Devon, who graduated from Falmouth University with a BA in Illustration. She likes to experiment with collage and texture to create her work and takes inspiration from her love of the outdoors and the beach.